Agriculture's Role in K-12 Education

Proceedings of a Forum on the
National Science Education Standards

Board on Agriculture

Professional Scientific Societies Related to
Agriculture, Food, and the Environment

National Research Council

NATIONAL ACADEMY PRESS
Washington, D.C. 1998

NATIONAL ACADEMY PRESS • 2101 Constitution Avenue, N.W. • Washington, DC 20418

NOTICE: The project that is the subject of this report was approved by the Governing Board of the National Research Council, whose members are drawn from the councils of the National Academy of Sciences, the National Academy of Engineering, and the Institute of Medicine. The members of the committee responsible for the report were chosen for their special competences and with regard for appropriate balance. In preparing its report, the committee invited people with different perspectives to present their views. Such invitation does not imply endorsement of those views.

This report has been reviewed by a group other than the authors according to procedures approved by a Report Review Committee consisting of members of the National Academy of Sciences, the National Academy of Engineering, and the Institute of Medicine. This report has been prepared with funds provided by the W. K. Kellogg Foundation, Farm Foundation, and the National Research Council.

The National Academy of Sciences is a private, nonprofit, self-perpetuating society of distinguished scholars engaged in scientific and engineering research, dedicated to the furtherance of science and technology and to their use for the general welfare. Upon the authority of the charter granted to it by the Congress in 1863, the Academy has a mandate that requires it to advise the federal government on scientific and technical matters. Dr. Bruce Alberts is president of the National Academy of Sciences.

The National Academy of Engineering was established in 1964, under the charter of the National Academy of Sciences, as a parallel organization of outstanding engineers. It is autonomous in its administration and in the selection of its members, sharing with the National Academy of Sciences the responsibility for advising the federal government. The National Academy of Engineering also sponsors engineering programs aimed at meeting national needs, encourages education and research, and recognizes the superior achievements of engineers. Dr. William A. Wulf is president of the National Academy of Engineering.

The Institute of Medicine was established in 1970 by the National Academy of Sciences to secure the services of eminent members of appropriate professions in the examination of policy matters pertaining to the health of the public. The Institute acts under the responsibility given to the National Academy of Sciences by its congressional charter to be an adviser to the federal government and, upon its own initiative, to identify issues of medical care, research, and education. Dr. Kenneth I. Shine is president of the Institute of Medicine.

The National Research Council was organized by the National Academy of Sciences in 1916 to associate the broad community of science and technology with the Academy's purposes of furthering knowledge and advising the federal government. Functioning in accordance with general policies determined by the Academy, the Council has become the principal operating agency of both the National Academy of Sciences and the National Academy of Engineering in providing services to the government, the public, and the scientific and engineering communities. The Council is administered jointly by both Academies and the Institute of Medicine. Dr. Bruce Alberts and Dr. William A. Wulf are chairman and vice-chairman, respectively, of the National Research Council.

Library of Congress Catalog Card Number: 98-84942

International Standard Book Number: 0-309-06048-6

Printed in the United States of America

STEERING COMMITTEE ON AGRICULTURE'S ROLE IN K-12 EDUCATION

CONRAD J. WEISER, *Chair*, College of Agricultural Sciences, Oregon State University, Corvallis

C. LEE CAMPBELL, Department of Plant Pathology, North Carolina State University, Raleigh

ROLAND OTTO, Center for Science and Engineering Education, Lawrence Berkeley National Laboratory, Berkeley, California

DEBORAH NEHER, Department of Biology, University of Toledo, Toledo, Ohio

ROBERT ZIMBELMAN, Executive Vice President, American Society of Animal Science and Federation of American Societies of Food Animal Sciences, Bethesda, Maryland

---

*Through December 1997.
†Through December 1996.
‡Through April 1996.
§Through June 1996.

Preface

The Board on Agriculture organized a Forum on Agriculture's Role in K-12 Education to provide an opportunity for agricultural professional societies to explore ways that examples from agriculture, food, and environment systems can be used to enhance inquiry-based science education. Participants discussed how professional societies could enhance continued education of K-12 teachers, improve school science programs, and increase collaborations with other professional societies and science teachers.

William DeLauder, Board on Agriculture member and president of Delaware State University, convened the forum of 71 scientists and educators representing 42 professional societies. Harold Pratt of the National Research Council walked participants through the *National Science Education Standards*. Pratt compared the *National Science Education Standards* to a national road map for achieving scientific literacy. Paul Williams, plant pathologist from The University of Wisconsin, made the keynote presentation. He urged participants to nurture and sustain the natural curiosity of students for science. Williams demonstrated inquiry-based approaches to teaching and learning biological science with Wisconsin Fast Plants™ (fast-growing *Brassica* plants).

Next a panel of educators and scientists exchanged views on the following issues: 1) curriculum decisions and design; 2) teacher education (pre-service, in-service, and other professional development opportunities for teachers; 3) educational resources; and 4) community support for inquiry-based science education. Perspectives were provided by Patricia Hoben, research director of Minnesota Public Utilities Commission in Minneapolis; Michael Klentschy, school superintendent from El Centro, California; Kathy Scoggin, early-education teacher from Minneapolis, Minnesota; and Judith Williams, high school biology teacher from

Central City, Nebraska. Jan Tuomi, director of the National Research Council's Regional Initiatives in Science Education, moderated the discussion. Panelists mentioned that scientists can help teachers understand inquiry-based learning processes. Participants stressed scientist-teacher partnerships as a way for scientists to become involved in K-12 science education reform.

Scientists and educators reconvened into smaller groups to discuss the *National Science Education Standards* and professional society roles in K-12 science education. Two questions were used as a basis of discussion. In Question 1, societies were asked to share their experiences in undertaking education-related activities or initiatives, identify those which were most effective, and make recommendations for other societies to consider. Their responses indicated that most societies are involved in K-12 education, but the level of involvement varies considerably. In Question 2, participants were asked to identify new directions which societies could explore, based on their understanding of the *National Science Education Standards* and the educator panel discussion. Scientists and educators indicated that professional societies need to enhance communications and collaborations among professional societies, educators, universities, and local communities.

William DeLauder synthesized the presentations and discussions made over the one and one-half day conference. DeLauder urged scientists to take responsibility for helping to improve education for all children. He reminded them that scientific society members can make a difference in science education reform by becoming involved with their professional societies at a national level and as individuals within their local communities.

Forum participants identified many ways that scientists can improve science education for K-12 students, undergraduates, and future teachers attending our colleges and universities. The proceedings contained herein do not include any recommendations but are intended to reflect the perspectives of forum participants. We hope the proceedings will stimulate further interactions between teachers and scientists and continue the momentum generated by this national meeting.

CONRAD WEISER, *Chair*
Forum Steering Committee

Table of Contents

APPENDIXES

Boxes and Figures

BOXES

FIGURE

Overview

Throughout America, school systems are involved in K-12 reform efforts to improve teaching and learning of all children. Because the *National Science Education Standards* call for dramatic changes in teaching and learning, many individuals will need to work together to realize a vision of scientific literacy. Professional societies and their members can make an important contribution to the improvement of K-12 science education.

Scientists recognize the importance of being educators, but too often teachers are expected to carry full responsibility for educating K-12 students. Some agricultural scientists are concerned about the lack of attention to agriculture, food and the environment in the science-content descriptions of the *National Science Education Standards*. Teachers indicate that they want to learn more about investigative processes used routinely by researchers. Although scientists are becoming involved in reform efforts, scientists and educators acknowledge that more could be done to enhance their collaboration. Those issues and more related to science reform need clarification. To facilitate that process, the National Research Council convened a forum of educators and professional societies broadly associated with food, agriculture, and the environment to exchange views on the role of scientists in improving science education.

Scientist-teacher partnership building can be an effective mechanism for making improvements in science education. Because teachers and scientists come from two widely different cultures, this divergence can inhibit mutual contact. Nevertheless, scientist-educator collaborations should be encouraged for reform efforts. Scientists can provide teachers with an understanding of scientific inquiry and content knowledge; teachers can help scientists understand learning processes and diversity of students. Participants suggested that partnership building is a means to support long-term interest in science education reform.

There was some discussion that agricultural and food-system literacy is essential and should be promoted by the societies. Agricultural sciences address not only production agriculture but the whole food and fiber system (from farm to table). Food is a common denominator for all children and is a useful way to get children's attention about agriculture. In developing agricultural examples for curriculum materials, collaborators should incorporate basic learning outcomes defined by the content standards.

While research is important to gain new knowledge, scientists have an obligation to make their knowledge accessible to the educational community. The skills of being a good communicator need to be promoted along with those of being a good scientist and a good technician. Agricultural scientists can help the public appreciate science by communicating with students and teachers using clear and understandable language. Scientists can enhance education in K-12 science classrooms through their local school district and can become more connected to their communities. Through scientist-teacher collaborations, scientists can:

- serve as a resource for content for teachers;
- rethink science fairs and classroom assessment;
- provide extra hands in classroom demonstrations;
- connect the teacher to the world of professional science;
- communicate your excitement about science;
- model science as inquiry; and
- foster scientific collegial interactions with educators.

Scientists also can participate in K-12 science education reform through their professional societies. Forum participants can:

- inform their staffs and other societies of the *National Science Education Standards*;
- increase the visibility of K-12 education among their members and at annual society meetings;
- invite K-12 educators to society meetings;
- obtain joint membership in K-12 educators' societies; and
- recognize and award members for involvement in K-12 education.

Participants recognized that coordinated efforts among disciplines and across societies could enhance outreach efforts. The need for an umbrella group to improve collaboration among the various societies was discussed in-depth. Participants suggested that a coordinating mechanism could increase the impact on K-12 science education.

The implications of the science standards on higher education are potentially enormous. Even though K-12 education was the focus of this forum, it was suggested that reforms in K-16 education should also be discussed. Today's high school students are tomorrow's college students. Participants questioned whether undergraduate instructors will be able to meet the needs of the new generation of high school students trained under the new *National Science Education Standards.*

Incentives will be needed to encourage scientists to get involved in reform efforts in K-12 classrooms. Although scientists might be inspired to get involved in service activities, many institutions take a dim view of this. Requirements for promotion and tenure provide disincentives for these interested scientists to give the time required to serve the reform effort. Therefore, incentives need to be considered at the university level. As members of professional societies and faculties, scientists need to communicate the importance of educational reform to other faculty members, department heads, and deans. A new era in education is beginning, and business cannot proceed as usual if scientists expect to have an impact. Scientific societies have performed a broad range of services in higher education, adult learning, and provision of technologies. Professional societies must now think about laying the foundation of early student learning in science and emphasize K-12 education.

Introduction

The education of future generations may be the single most important challenge that this nation faces. Traditionally, society has placed the burden of kindergarten through twelfth-grade science education on teachers. However, educational systems are changing, and schools and communities are joining together to implement systemwide reforms in science education.

Public and professional support for national science education standards was a catalyst for change. In 1991, a group of science education associations asked the National Research Council (NRC) to coordinate development of national standards for content, teaching, and assessment for K-12 science education. In early November 1994, the first full draft of these standards was publicly released for a national review. Approximately 40,000 copies of the draft were disseminated to more than 140 focus groups, including state departments of education, local school districts, and professional societies across the country. Contributions from focus groups were incorporated into the final draft, which was released as the NRC's consensus report, *National Science Education Standards* (1995). The report provides a focus for discussions of reform among groups with a variety of interests.

In the agricultural sector, shifts toward more diverse consumer-oriented food markets, competition for global trade, and increased safeguards for the environment are putting greater demands on agriculture's base in science and technology. Agriculture is no longer only about production, but also involves the entire food and fiber system. Universities and colleges need to continue to attract qualified students into the agricultural sciences. To do that, promotion of agricultural concepts within a framework of food and environmental systems would appeal to students and teachers. However, many scientists who have spent their entire careers in research laboratories feel ill equipped to work with K-12 students.

Educators in K-12 classrooms are experienced with child development and pedagogy but know little about using inquiry—a key strategy for learning science. Scientists have traditionally provided advice on science content, but have not adequately communicated methods of inquiry. A question is raised as to appropriate roles for scientists in K-12 science education.

Until recently, science education reform has been occurring at local, state, and national levels with little focus on the links between these levels. Although a wealth of experience is represented by the agricultural professional societies, increased specialization has created barriers across disciplines and with educators. This lack of integration has led academic scientists, professional societies, and the K-12 education community to view educational reform as their own separate agendas. As a result, teachers and scientists have little contact with each other. Without coordinated and sustaining support for teachers, the education of our nation's children could be at stake.

Because of these concerns, strong interest developed in providing a forum where educators and professional societies could discuss the roles scientists can play to improve K-12 science education. On November 17-18, 1995, the National Research Council's Board on Agriculture and Center for Science, Mathematics, and Engineering Education convened a group of scientists and educators representing 42 professional societies to discuss ways in which examples from the agriculture, food, and environmental system can be used to enhance teaching of inquiry-based science. Educators shared their experiences in implementing the *National Science Education Standards* and provided their perspectives on teacher education, curriculum decisions and design, coordinating community resources, and sustaining support for science education reform. Scientists discussed their roles in undertaking education-related activities and new directions professional societies might take to make an impact on the quality of science education. The following proceedings summarize issues raised by forum participants and ideas to enhance collaboration between scientists and educators. The summary is limited to the views and opinions of those participating in the event.

A NATIONAL VISION

Students should learn science for many reasons, and there are a number of advantages and outcomes if they do. Scientists agreed that children should understand and appreciate the richness of the world they live in—a world that they should and often do get excited about. "Those of us who are engaged in science know what a rewarding venue it is for our own creativity," commented Paul Williams, plant pathologist at the University of Wisconsin and inventor of Fast Plants. "A challenge for us is to determine how best to convey the nature of science as we understand it to the generation that follows."

Nurturing an interest in science is not a new concept. Paul Williams reminded the audience of 70 participants that, "It isn't as if this responsibility suddenly arrived with the *National Science Education Standards*—the task has always been before us. We all know how precious and fragile the human characteristic of curiosity is. As parents we know that in young children curiosity is pristine, ever present, and unblemished. Our challenge as scientists concerned with education is how best to nurture and sustain this most valued human characteristic in our educational system through teachers and ensure that they experience the pleasures of science as we do."

The professional societies represented at the forum consisted of scientists from highly specialized, agriculturally oriented disciplines. Each of the disciplines has adopted a set of highly specialized terminology to enhance communication. Paul Williams explained the limitations of that strategy. "This terminology essentially constitutes the secret language of science that isolates scientists from the public, teachers, and children. One of our primary responsibilities is to translate, into clear common language, the secret language of our diverse disciplines, so that we and they may engage in the exciting discourse that science will generate."

There is a strong argument for a scientifically literate population, explained Harold Pratt, director of the K-12 Policy and Practice Division of the National Research Council's Center for Science, Mathematics, and Engineering Education. "A greater understanding of science can lead to a higher quality of life. Scientific principles can be applied to personal decision making and to discussions of scientific issues that affect society. Our nation's productivity will benefit from a work force well grounded in skills of science and technology."

THE NATIONAL SCIENCE EDUCATION STANDARDS

The standards provide criteria for examining where education is today and a roadmap to where the nation wants education to be tomorrow. From a science education perspective, national standards have multiple meanings and multiple uses. First and foremost, they identify what students should know and be able to do at various grade levels and ages. Second, they specify the support needed for teachers and students: the resources, facilities, planning time, and opportunities for teachers to participate in decision-making processes. Beyond the classroom, the national standards describe program and system standards.

At the forum, Harold Pratt guided participants through the *National Science Education Standards* (1996), concentrating on science content because that area is of great interest to scientists.

The following four principles guided the development of the *National Science Education Standards*:

1. Science is for all students.
2. Science education should reflect the intellectual and cultural traditions of contemporary science.
3. Learning science is an active, inquiry-based process.
4. Improving science education is a part of a systemwide reform in education.

BOX 1
An Overview of the *National Science Education Standards*

The *National Science Education Standards* defines the science content that all students should know and be able to do. Standards are provided for content, teaching, assessment, professional development, programs, and systems. They are based on the premise that learning science is something students do, not something that is done to them. They envision an active learning process in which students describe objects and events, ask questions, formulate explanations, test those explanations, and communicate their ideas to others. In this way, students build strong knowledge of science content, apply that knowledge to new problems, learn how to communicate clearly, and build critical and logical thinking skills.

Content Standards

The content standards describe the knowledge and abilities that K-12 students need to become scientifically literate. Scientific literacy is the knowledge of scientific concepts and processes required for personal decision-making, participation in civic and cultural affairs, and economic productivity. The content standards are divided into eight categories:

1. Unifying concepts and processes
2. Science as inquiry
3. Physical science
4. Life science
5. Earth and space science
6. Science and technology
7. Science in personal and social perspectives
8. History and nature of science

The first category of the standards, unifying concepts and processes, identifies powerful ideas that are basic to the science disciplines and help students of all ages to understand the natural world. This category is presented for all grade levels because the ideas are developed throughout a student's education. The other content categories are clustered for grades K-4, 5-8, and 9-12. Students develop knowledge and abilities in inquiry, which is the basis of learning in physical, life, earth, and space sciences. The science and technology category links the natural and designed worlds. The category of personal and social perspectives helps students to observe the personal and social impact of science and to develop decision-making skills. The history and nature of science category helps students to perceive science as a human experience that is ongoing and changing.

Teaching Standards

Science teaching is central to the vision of science education presented in the *Standards*. Teachers of science have theoretical and prac-

continued

BOX 1—Continued

tical knowledge about learning, science, and science teaching. The teaching standards describe the skills and knowledge teachers must have to teach science well.

Teachers are encouraged by the *National Science Education Standards* to give less emphasis to fact-based programs and greater emphasis to inquiry-based programs with in-depth study of fewer topics. However, to attain the science education described in the *National Science Education Standards*, more areas than teaching practices and materials must change. The routines, rewards, structures, and expectations of districts, schools, and other parts of the school system must be used to endorse the vision, and teachers must be provided with resources, time, and opportunities to change their practice. Teachers can use the program and system standards to communicate that need to administrators and parents.

Assessment Standards

The assessment standards provide criteria to judge the progress of scientific literacy throughout the educational system. They can be used in preparing evaluations of students, teachers, programs, and policies. The assessment standards identify the essential characteristics of effective assessment policies, practices, and tasks. Teachers who use the assessment standards will think differently about what and where to assess, and the best way to determine what their students are learning. They will consider their students' fundamental understanding, their progress in developing understanding, and their need for alternative ways to demonstrate what they know.

Professional Development Standards

The professional development standards make the case that becoming an effective teacher of science is a continuous process, stretching from preservice throughout one's professional career. The professional development standards can be used to help teachers of K-12 science to have ongoing, in-depth learning opportunities similar to those required by other professionals. Professional development standards call for teachers to have the following opportunities:

- Learn science through inquiry.
- Integrate knowledge of science, learning, and teaching.

BOX 1—Continued

- Engage in continuous reflection and improvement.
- Build coherent, coordinated programs for professional learning.

Program Standards

The program standards address the need for comprehensive and coordinated experience in science across grade levels and support for teachers so that all students have the opportunity to learn. The program standards will help schools and districts translate the *National Science Education Standards* into effective programs that reflect local contexts and policies. Effective science programs are designed to consider and draw consistency from the content, teaching, and assessment standards, as well as the professional development, program, and system standards.

System Standards

The system standards call on all parts of the educational system—including local districts, state education departments, and the federal educational system—to coordinate their efforts and build on one another's strengths. The system standards can serve as criteria for judging the effectiveness of components of the system responsible for providing schools with necessary financial and intellectual resources.

The Road Ahead

The changes required to achieve the vision of the *National Science Education Standards* are substantial and will continue well into the 21st century. No one group can implement them. The challenge of a *National Science Education Standards*-based science program extends to everyone within the educational community. Change will occur locally, and differences in individuals, schools, and communities will result in different ways to improve the system, different rates of progress, and different school science programs. What is important is that change be pervasive and sustainable, leading to high-quality science education for all students.

SOURCE: National Research Council. 1997. *Introducing the National Science Education Standards.* National Academy Press: Washington, D.C. For ordering information on *National Science Education Standards* or *Introducing the National Science Education Standards*, call the National Academy Press Bookstore at (800) 624-6242 or (202) 334-3313 or browse the National Research Council's World Wide Web site at: http://www.nap.edu/.

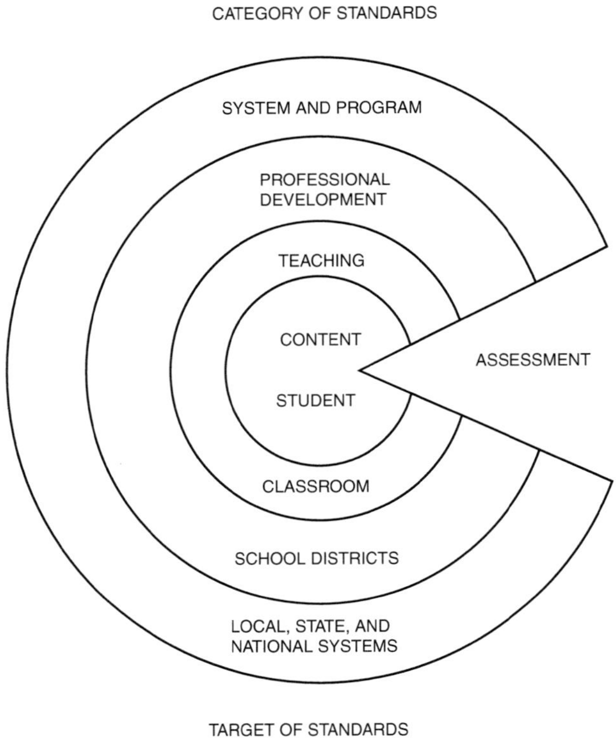

FIGURE 1 A systemic view of the *National Science Education Standards*. Source: Harold Pratt, National Research Council.

The *National Science Education Standards* report describes content as the specific capacities, understandings, and abilities in science, whereas curriculum is the structure, organization, balance, and presentation of the content in the classroom.

The eight broad categories of science-content standards for K-12 (listed in Box 1, above) represent a wider definition of content than was used in the past and include the basic principle of learning science through inquiry. According to the *National Science Education Standards*, inquiry "requires that students combine processes and scientific knowledge as they use scientific reasoning and critical thinking to develop their understanding of science." The intention of the *National Science Education Standards* is to describe "the abilities to carry out an investigation, as well as the understanding of the nature of the inquiry and experimental processes."

Scientists were eager to discuss the content standards, and clearly some were concerned that the *National Science Education Standards* contained few references to food, fiber, and renewable resources. Vernon Cardwell, professor of agronomy at the University of Minnesota, asked for help in identifying opportunities where examples from agriculture could address the content areas of the *National Science Education Standards*. Understandably, many educators and scientists might have difficulty reconciling the interdisciplinary aspects of agriculture with the basic science categories listed in the *National Science Education Standards*. In his reply, Harold Pratt explained that the national standards were not designed to prescribe a particular curriculum or course of study. "If you look at the program standards, you see they advocate a variety of contexts for developing curriculum materials and teaching strategies. The curriculum can take many forms; the instruction can blend things together in a variety of ways. We simply needed a way to catalog outcomes—list the knowledge that we wanted students to achieve."

According to the *National Science Education Standards*, assessment standards "provide criteria to judge progress toward the science education vision of scientific literacy for all. The standards describe the quality of assessment practices used by teachers and state and federal agencies to measure student achievement and the opportunity provided students to learn science." To this end, the *National Science Education Standards* incorporates assessment strategies as a feedback mechanism to measure the progress of reform throughout the educational system (see Figure 1).

If scientists are going to be effective in reform efforts, they need to understand that science education is much larger than content. It also includes instructional strategies for inquiry-based science education, professional development opportunities for educators, sufficient financial and intellectual resources; and assessment strategies. Also important to the educational system are universities and colleges; science museums; science resource institutions; and parents. Addressing what students should learn without thinking about the total context or system in which they learn is short sighted. The *National Science Education Standards* provides a framework to unify all the interacting components needed to improve science education in the long term.

A Scientist's Role in K-12 Education

Implementation of the national science education standards will require major changes in the educational system. Participation in the reform effort must be multifaceted to reflect the multiple dimensions of the standards, and professional societies and scientists are uniquely qualified to contribute. Scientists can determine their own level and form of participation on the basis of their time and expertise, but every scientist has something to offer. The goal is to ensure that each student is offered an equal opportunity to learn science.

Scientists play an important role as science instructors in university and college classrooms. They also can assist in reform efforts through membership in their professional societies. Scientists could consider becoming involved in the K-12 science classroom. In doing so, scientists will become more connected to their communities, and children will benefit.

COLLEGE AND UNIVERSITY CLASSROOMS

The *National Science Education Standards* recognizes that college and university faculty have a direct influence on changing pedagogical practices in the K-12 classroom. "Most of us in this room are from academic institutions of higher learning where undergraduates are being trained to become our next teachers. It is with these undergraduates that our responsibility for teaching science must lie. Providing future teachers with an understanding of science should not be the responsibility of the colleges of education with expertise largely within the realms of pedagogy and human development." Paul Williams urged scientists to take a more active approach to science education. "These future science teachers need formative experiences found only in your science laboratories and classrooms. These students need to be mentored with particular care and understanding so that when they graduate, they will carry with them the special insights that only you will be able to pass on to them."

The National Science Education Standards describe an educational system to achieve a scientifically literate population. They impact student learning outcomes at all levels. At the center is the student target and the science education content category. Assessment can be viewed as a primary feedback mechanism in the science education system that cuts across the educational system hierarchy from the student to national levels.

"Science is too often taught as natural history in this country," Paul went on to say. "Natural history is the catalog and narrative of natural phenomena that other people observed and discovered. Our challenge has to be to take what we find exciting and not just treat it as history—don't just tell them who discovered this; let them discover it for themselves, and they will be the possessors and owners and understanders of it. To that end, I actively use the pedagogical stratagem of reserving the right *not* to answer a good question from a student. But if it's a good question, why won't I answer it? Because good questions lead students and teachers to engage in the process that we call science. And that of course will lead them to another set of questions, which is what science is all about.

"Let me suggest that in order for us to better understand how to help teachers engage in science in their classrooms, that we examine doing more science and less natural history in our own undergraduate college and university classrooms. There are now available many resources to assist us in changing the way we teach in our own classes. By improving the ways we teach science in our own classes, we will be making an impact on the way science is taught in K-12 classrooms, because the teachers of the future are among our undergraduate students today."

In his concluding remarks, Paul Williams asked scientists to consider the potential impact of students taught in high school under the new standards on university and college classrooms. "It's interesting that we're discussing the national science standards at this conference because indeed they are centered on precollege education. That's appropriate. But do you know from where the tidal wave is going to come? It's going to come from the bottom up. Will we be prepared to readjust the way we teach science at the college level to meet the need of the new generation of high school students trained under the *National Science Education Standards*? How will we realign what we teach to meet the expectations of these students?"

PROFESSIONAL SOCIETIES

Scientists are getting involved in K-12 reform initiatives through their professional societies. The forum noted that the level of support for these initiatives varies with the mission and size of the societies. For instance, the American

Chemical Society is a broad-based society reaching out to over 130,000 members; others might have fewer than 500 members. In particular, society representatives identified the need for dedicated staff to increase their participation in education-related activities.

Participants identified many examples of society activities that focus on K-12 education:

- summer internship programs for teachers and students;
- development of lesson plans and choosing textbooks with teachers;
- summer field programs for high-school students;
- development of video materials promoting science literacy;
- writing contests to encourage student involvement in science;
- scientists serving as consultants or demonstrators to schools;
- society sponsorship of science fairs;
- educator participation in professional society meetings;
- scientist participation in design of curriculum materials.

Educators expressed concern that they are unfamiliar with K-12 activities of the professional societies. In some cases, teachers are not sure how to access information from the societies. At the other end of the spectrum, teachers are overwhelmed with offers from scientists. Participants agreed that professional society members should work harder to coordinate their efforts to aid teachers.

Forum participants identified the following examples of ways members of professional societies can become involved:

- Inform their staffs and other societies of the *National Science Education Standards*.
- Increase the visibility of K-12 education among their members and at annual society meetings.
- Invite K-12 educators to society meetings.
- Obtain joint membership in K-12 educators' societies.
- Recognize and provide awards to their members for involvement in K-12 education.
- Recognize that coordinated efforts among disciplines and across societies could enhance their outreach efforts.

K-12 CLASSROOMS

Most educators recognize that K-12 classrooms are unfamiliar settings for academic scientists. Working in a K-12 classroom will not be easy or appropriate for all scientists. Scientists will need to be properly prepared and trained to work in a K-12 classroom. By working directly with students, scientists can be role models for students and encourage them in science.

Training and professional opportunities are needed for K-12 teachers, as well as for prospective teachers. Although teachers are central to educational reform, they cannot implement the *National Science Education Standards* without help. Teachers need an understanding of inquiry-based education and continuing opportunities for professional development. Because of their expertise in scientific research, scientists have an opportunity to assist teachers in keeping current on scientific concepts, understanding the investigation process, and practicing solving real problems.

Jan Tuomi, director of National Research Council's Regional Initiatives in Science Education Program encouraged scientists to become involved in K-12 education. "If you want to make a high impact on the quality of K-12 science education, then you've already taken the first step, and you're ready to take the next, which is asking, 'What can I do? How do I get started? What works? What's a good way or a not-so-good way to approach this?"

Scientists might have expectations of what happens in a K-12 classroom based on their own school experiences or what they have experienced in their laboratories. Teachers, on the other hand, can provide a real world perspective that may be unfamiliar to scientists. Some educators at the forum suggested that scientists visit a K-12 science class. By observing, scientists can gain insights to the complexity of the teaching and learning environment.

The *National Science Education Standards* recognizes that children will achieve different degrees of understanding of science, depending on their interests and abilities. Scientists need to be aware of the context in which children are learning and the things going on in their lives. Kathy Scoggin, a Minneapolis teacher, described a real situation: "There are kids that come to school that haven't had breakfast, their mom had breast cancer surgery yesterday, their dad's in jail, two others are homeless, and they're coming from a shelter. That's what it can be like. But school can be a safe haven for them. Despite these problems, kids want to learn and get very excited about science. It's important to understand what the situation is out there today."

That situation might cause uneasiness in faculty members who are accustomed to work environments far removed from K-12 classrooms. Michael Klentschy, superintendent of El Centro schools in California, tried to put that fear into perspective. "Just as you feel comfortable working in your research laboratory, our teachers feel comfortable dealing with the social issues connected to teaching and learning. That's where the cooperation between scientist and teacher is really very important, so that we are breaking down barriers that might seem frightening to both of us."

Educators at the forum indicated that teachers go into education for many reasons and have varying levels of science in their backgrounds. Kathy Scoggin, for instance, grew up on a farm and developed an early appreciation for science and the natural world. According to Michael Klentschy, many elementary teachers are women who were not encouraged to excel in science when they were in school or when they were Liberal Arts majors with little exposure to science. Because teachers have different levels of teaching experience, their requirements from the science community will vary. Kathy Scoggin described different ways scientists can interact: "New teachers might have 9-12 units of science and one science-methods class. Scientists can help them by teaching them some basic scientific concepts. Honestly, these new teachers are trying to figure out very basic skills like how to get students to stay seated. The second group [is composed of] teachers who have between 5 and 20 years of experience. Many have built a strong base of scientific knowledge over the years and they are well versed in pedagogy. Scientists can be of help by showing these teachers how to incorporate the *National Science Education Standards* into their teaching practices. Finally, there are those teachers who are going to retire in the next 5 years. Scientists should try to motivate these teachers by showing them new things that can make their final years of teaching really productive."

There are many ways scientists can become involved in improving science education in K-12 classrooms. Some suggestions are listed below:

- Serve as a resource for course content for teachers.
- Collaboratively rethink science fairs and classroom assessment.
- Provide extra hands in classroom demonstrations.
- Connect the teacher to the world of professional science.
- Communicate your excitement about science.
- Consistently model science as inquiry.
- Foster collegial interactions with educators.

Scientist and Teacher Partnerships

Participants agreed that cooperative endeavors between scientists and teachers are an excellent strategy for teacher-education training. Scientist-teacher collaborations can be used to develop effective curriculum materials and implement inquiry-based teaching methods. Michael Klentschy suggested that the best types of partnerships consist of a scientist and a lead teacher working together to facilitate a group of teachers who have either a common interest or taught a common grade level. The next section describes two examples of successful scientist-teacher partnerships.

PIONEERS IN SCIENTIST-TEACHER PARTNERSHIPS

Michael Klentschy
Superintendent, El Centro School District, El Centro, California

Michael Klentschy described his partnering experiences in his former district of Pasadena, California—an urban area where 85 percent of about 22,000 schoolchildren come from the two lowest census-track income categories. "Though at the beginning it wasn't one of my highest priorities for action, science education reform became one when two faculty members from California Institute of Technology came and visited me in my office. I thought they were high-school teachers, but they happened to be two parents. These two gentlemen really got the ball rolling with science education reform in Pasadena." That partnership is a model for effective teacher-scientist collaborations now in place in many school districts in the United States.

"At that critical point when the science community began working with the educators, the first thing we had to do was learn each other's language. You have little codes and acronyms; we do, too. It's kind of a toss up as to who has the most. When we started speaking in plain English, we found that we had more common ground than we had differences. Scientists play a critical role because they bring content background to the table that we don't have, and educators can bring classroom experience to the table that scientists don't have. What we do is develop a common experience for each other that ultimately we pass on to the next generation of students. It's kind of a variation on the Fast Plants ™ experiment in terms of opening up and planting those little pods. That's basically the relationship that we can establish as educators and professional scientists. It's that planting and replanting that is going to make the difference in the types of curriculum decisions we make in the future."

A SCIENTIST AND TEACHER MAKE BROAD, UNEXPECTED IMPACTS

Roland Otto
Head, Center for Science and Engineering Education, Lawrence Berkeley National Laboratory

"At Lawrence Berkeley National Laboratory we have had fourteen years of experience with high school and middle school teachers working as research associates in small groups with scientists, technicians, graduate and undergraduate students. Since 1983 over 270 appointments have been made in every field of scientific research and development at the Laboratory. These summer research assignments are reported by some teachers to be the most significant professional development experience of their careers. For others the insight into the world of science is invaluable providing them with renewal, revitalization and recognition that allows them to contribute to the reform in science education. At the same time the impact that these teachers have had on the scientists with whom they have worked and interacted has been significant. Teachers bring to the scientists insights into today's classrooms and the day-to-day challenges faced in educating today's students. These insights have motivated many scientists to greater and more meaningful involvement in K-12 education through long term partnerships with teachers.

"In 1989 Dean Rockwell, a biology teacher from Macomb, Illinois was one of about 12 teachers selected for the LBL summer program. For some reason we had some trouble placing Dean in a life sciences laboratory so we placed him in an unlikely spot, with a physicist, Tony Hansen, in the atmospheric aerosol research group. Tony was the inventor of an instrument capable of measuring graphitic carbon aerosol (soot), a primary atmospheric pollutant from combustion of fossil fuels (Tony's patented device has been used to establish the existence of Arctic Haze and to track the course of Kuwait oil fires after Desert Storm). After learning more about the instrument, Dean discussed the possibility of doing soot measurement experiments with his students. The challenge they faced, given Dean's science budget, was that they would have to develop a procedure and instrument for under $10. Tony was the kind of inventor who preferred to use ordinary everyday materials, which he often gathered in his garage. When he had filled up one garage he rented another. Tony gladly took on the challenge and by the end of the summer he and Dean had developed a procedure that used Kleenex, a vacuum cleaner, a large yard bag, a light bulb, plastic cups, and a $2.40 photo cell that could be attached to a voltmeter. To Tony's surprise this low-tech procedure produced data that had a correlation coefficient of .999 when compared with data from measurements using his best instrument. The procedure and more recent improvements that include using an aquarium pump for longer time sampling is being published in the Journal of Chemical Education.

"The story continues. Tony visited the Soviet Union shortly after the Berlin Wall fell. The air pollution problem, particularly the soot concentration in the Eastern European countries can be 10 times the concentration on a bad day in Los Angeles. The scientists there discussed their limited resources and their desire to set up a network to monitor the atmospheric soot concentrations. Tony suggested to his Eastern European colleagues that they use the system he and Dean Rockwell had developed. This idea was picked up by scientists in Slovenia with whom Tony had been working. After starting the project in Slovenia's schools Dean was asked in 1992 to go to Slovenia and train elementary and high school teachers to make these measurements. Dean did this and at the same time received recognition from the country's leaders for his contribution. The Slovenian national network was established in 1991, monitored by the National Institute of Chemistry and the results were reported in an international conference in Vienna in 1992. The report generated considerable interest from scientists in other Eastern European countries. As a result of the conference, the nation of Estonia, formerly a part of the Soviet Union, created a national network in 12 high schools using the Slovenia model, with financial support from the Soros Foundation. Dean has been invited back to Estonia this spring for lectures and tours of the sampling sites. A colleague of Tony's, Mirko Bizjak from the Hydrometeorological Institute of Slovenia has described this work in a Bulletin of World Meteorological Organization (Vol. 43, No. 1 January 1994, pp 60).

"Scientists have connections around the world that teachers do not. At the same time teachers know what works and motivate learning science in the classroom. By working in partnership in the scientists environment meaningful resources are developed for the classroom. The other important message is that children can participate in science discovery. For many teachers and students science is the cleaned up and simplified science in the text books and labs they do. There is often no concrete connection to the natural world they live and play in. In the story above, students of all grade levels connected to an open ended problem with the possibility of discovery. They became part of an investigation that was of interest to many others including real scientists. This level of deep connection with science should be an important goal for every teacher of science. It is the responsibility of the science community to partner with science teachers so this goal can be achieved."

SCIENTISTS TEACHING INQUIRY SKILLS

According to the *National Science Education Standards* (1996), inquiry-based learning will shape the future of science education. The *National Science Education Standards* describe, inquiry as "a step beyond science as a process, in which students develop skills, such as observation, inference, and experimentation. The new vision presented by the *National Science Education Standards* includes the processes of science and requires that students combine processes and scientific knowledge as they use scientific reasoning and critical thinking to develop their understanding of science." An inquiry-based process enables the student to learn how to design and interpret an investigation and communicate scientific ideas. Teachers guide students through discovery of real phenomena in classrooms, outdoors, and laboratory settings. Students begin to learn how to solve problems through teacher facilitation. Many educators believe that the greatest need of teachers is to learn and utilize inquiry-based approaches. Partnering with scientists should be a major part of this effort.

Kathy Scoggin, an elementary school teacher from Minneapolis, described her role. "Basically, it's a teacher's job to facilitate learning. So I cannot stand up there and tell my students, 'This is what you need to know today.' I need to find out my students' interests, what's going on in the world that I should expose them to; help them figure out their questions; and keep them challenged. That's what makes a positive learning experience."

Judith Williams, a high-school biology teacher in Central City, Nebraska, had a similar reaction. "I've heard so many people talk about scientists seeing their role as helping with content. I think you should help us learn how to do inquiry, so we can help our students solve problems. Most of us just don't know how to ask the right questions."

Scientists can help teachers implement the science education standards by modeling the inquiry process. "Teachers learn to teach. We learn about pedagogy, child development, a little math; but we don't learn about inquiry the way you do it," said Kathy Scoggin. "So when an education system is being set up to help us, that's probably one of the most valuable things to include; and if you look at the *National Science Education Standards*, they are organized around inquiry-based education."

Scientists can demonstrate the teacher's role as a facilitator by asking open-ended questions and showing them how to question the experimental variables, results, and conclusions. Michael Klentschy suggested that scientists could help teachers to keep the big picture in perspective. "Many times I have observed our teachers go through an activity-based approach to teaching science, but do so as a series of unconnected activities. Scientists can help teachers with questions like, 'What is this entire unit all about? How can we connect many of these activities?'"

Michael Klentschy suggested that scientists bring along their laboratory journals when they are in working groups with teachers. "As part of every unit in which I have been associated, all students maintain lab journals for record keeping and reflecting. When the scientists go through their lab journals, teachers often will say, 'Oh, you didn't get the right answer the first time.' The scientist may reply, 'We're still not sure what we're doing is the right answer. It's just a better-formed question.' That really helps break down some of the barriers."

Inquiry is about doing and learning by experience. Science teaching has evolved from the cookbook approach in which a laboratory exercise is done to reach a particular outcome to an experimental approach. Students engage in an open-ended investigation, find out what and why it happened, compare results with other students, and formulate predictions for next time. Kathy Scoggin added, "And if it doesn't come out the way you thought it was going to, you still have learned something. That's what it's all about. All of those things fit together. So, as far as it goes for creating good experiences, always keep in mind that it's got to involve doing and learning from what you are doing and it's building on that; it's scaffolding, that's what they call it in education."

SHARING SCIENCE RESOURCES

An effective science program has ample and appropriate resources to support inquiry-based learning. Educators and scientists agreed that there is a lack of sufficient suitable resources to facilitate implementation of the *National Science Education Standards*. For example, educators pointed to the need for library materials that incorporate the *National Science Education Standards*. Educators mentioned that CD-ROMs, television in the classroom, and interactive media are more appealing to students than traditional library books.

Much of the forum discussion centered on new opportunities created by increased use of the WorldWideWeb to access science education resources. Access to electronic media enable educators who are geographically isolated to acquire educational resources not generally available to them. Judith Williams, a teacher in Central City, Nebraska, indicated that she depends on electronic access to bulletin boards such as Genentech's *Access Excellence* for her high-school biology program. Society representatives suggested that societies could create WorldWideWeb home pages as a mechanism to communicate their K-12 educational activities. For example, the American Society of Microbiology has a home page on the WorldWideWeb that includes teacher-resource listings and an activities exchange. In addition, the Council of Agricultural Sciences and Technology (CAST) has a home page that can connect to 11 of their 31 membership societies, which also have home pages. Scientists anticipate that more societies will use the WorldWideWeb to post their materials and build communication with K-12 educators.

Society representatives identified the following examples of information sources available for K-12 teachers:

- The Soil and Water Conservation Society is presenting new materials on water, environment, and agriculture to be distributed through the Discovery Channel on television and CD-ROM.
- The Institute of Food Technologists has a variety of materials that it distributes free of charge, including videos on food science, brochures, and various types of booklets.
- The American Society for Horticulture has prepared videotape material that can be distributed readily.
- The American Society of Microbiology is producing a film to emphasize the societal benefits of microbes for bioremediation, decomposition, and fermentation.
- North Carolina State University is conducting workshops funded by the National Science Foundation that link scientists and high-school teachers (SCI-LINK). Teachers are brought into the scientist's laboratory and exposed to various methods, and then the teachers are given the opportunity to write curricula relevant to those projects.
- The American Society of Agronomy, Soil Science Society, and Crop Science Society of America have instituted a summer internship in Honduras for 12 students.
- The American Association for the Advancement of Science had a conference on agriculture and K-12 education.
- Representatives from the 1890 land-grant colleges host a biannual symposium and invite minority students. Papers and posters are presented and prizes are given. Attendance is about 400. Not only academic scientists, but also scientists representing industry are included in this endeavor.

Teachers continually need fresh ideas for their classroom activities. Because they are innovators, teachers will explore their local surroundings to find science-rich resources that will appeal to their students. For example, Michael Klentschy expects that the irrigated agricultural lands in his new school district of El Centro, California will provide a backdrop for numerous inquiry-based activities. Forum participants identified potential educational resources for teachers in rural areas:

- Land grant universities and colleges;
- U.S. Department of Agriculture (USDA) Natural Resource Conservation Service;
- U.S. Parks and Game Commission;

BOX 2

Impacts of the World Wide Web on Science Education

Paul Williams, University of Wisconsin

"At the Wisconsin Fast Plants™ Program, many people using the fast plants were contacting us for information. Our staff just could not cope with the overwhelming load of mail, newspaper costs, and telephone calls from teachers all over the country. We decided to put up a server on the World Wide Web for teachers, students, and researchers to share information and we could stay out of it. During the first week 40 people from around the world registered on our electronic mail site. We were disappointed because there was no interaction. It seemed that everyone was just sitting there waiting for someone else to start a discussion. Two weeks later the first message appeared. From where do you think that message originated? It came from a sixth grader in Kansas. A message popped up that said, 'Hi, I'm Joey, and I'm doing an experiment on the nutrition of fast-plants. Then Joey described his experiment and waited for a reply. Within a brief amount of time, a sixth grader from California responded to Joey. Since then, the electronic discussion has flourished.

Some thoughts struck me as a result of this incident: How am I going to be ready for those two sixth-graders when they arrive in my university in six or seven years? Can you imagine what science education will be for those two students when they reach middle school and high school? Will we be prepared to readjust the way we teach science at the college level to meet the need of the new generation of students trained under the standards? How will we change our teaching to meet their expectations?"

- Bureau of Land Management;
- Farmers and the Extension Service;
- Local utilities, including water and sewerage departments;
- U.S. Department of Defense bases; and
- Agricultural input suppliers.

DEVELOPING CURRICULUM MATERIALS

The *National Science Education Standards* provides criteria about the science content that should be used to guide the curriculum-development process. The content as translated into the classroom is usually determined at the state or district level, or the ad hoc curriculum is sometimes the table of contents in the course textbook. Educators will need curriculum materials aligned with the *National Science Education Standards* and community-wide assurances that these materials will serve the needs of students.

Teachers acknowledge the efforts of scientists to provide curriculum materials for science classrooms, but some of these resources are not adaptable to a particular teaching curriculum and might not appeal to students. As a result, many of these materials stay on the shelf and do not get into students' hands. "We can put out the most wonderful curriculum materials in the world, but when the classroom door closes, teachers do pretty much what they feel most comfortable doing," admitted Michael Klentschy. Judith Williams stated, "When it comes down to who makes the decision about what will be taught in my school—it's me. I'm the biology teacher and I teach every student biology!"

Kathy Scoggin suggested one way for scientists to get around that problem. "When you are asking yourself, what are the big concepts that I think they should know in food and fiber, consider how that connects to these kids' lives, because that's what they're going to learn." Richard Stuckey, Council for Agricultural Science and Technology, elaborated on this farm-to-table approach. "Food is a common denominator for all children, whereas agriculture by itself is not. Agriculture is a part of that food system, so I don't think that it delimits what can be included in the curriculum regarding agriculture. But I think you can get the attention of kids if you start talking about food and then look at that total system."

Agricultural applications can be used as examples to support outcomes described in the science-content categories. Harold Pratt emphasized the need to incorporate basic scientific principles into development of curriculum materials. For example, in the American Chemical Society's *Chemistry in the Community*, a contemporary problem is defined and then a solution is determined through application of fundamental scientific principles. In the same way, agricultural, food, and environmental topics can serve as a basis for learning basic scientific principles.

William DeLauder, president of Delaware State University, added that Delaware has established a science initiative that originated with a grant from the National Science Foundation. "Teachers developed a backpack experiment to facilitate learning of basic concepts in chemistry and physics because the backpack is a familiar object to youngsters. Children learn chemical principles through studies on chemicals and materials used to manufacture the backpack. Laws of physics can be applied to investigations of the backpack's strength and weight. The backpack is an example of a way to teach basic science concepts by using familiar objects."

Representatives from the professional societies stressed the importance of educator-scientist partnerships for curriculum development activities. Scientists can offer the content knowledge in this process, but teachers have a great deal to teach scientists about pedagogy. For example, in the El Centro California schools, at least two scientists in the area volunteer to be mentors for each of 32 science units. Michael Klentschy emphasized that these scientists do not sit in the laboratory and write the unit. "These are collaborative efforts between classroom teachers who are using them and the content scientists who help provide perspective on the larger ideas. Teacher involvement becomes even more important as we link these units together all the way through to the end of eighth grade."

Participants suggested that societies should seek new collaborative arrangements with teachers, scientists outside of their discipline, and other professional societies. Although some curriculum materials could be adapted to fit the new standards, in other cases collaborations could bring new ideas into the process. In addition, societies that share a discipline could choose to work together on a project. For instance, some of the smaller societies that have little infrastructure might benefit by linking their programs to those of larger societies with similar interests. Some professional society representatives admitted that conflicts do arise among societies, especially in regard to sharing credit. They cautioned that these potential problems should be addressed in advance.

Participants identified some cooperative curriculum projects under development. For example, Food, Land, and People is a coalition of stakeholders that is developing science-education materials that are being pilot tested by over 400 teachers.

Harold Pratt, of the National Research Council, advised participants that the approach of using the fiber and food industry for building curriculum materials can be powerful as long as efforts are coordinated with larger groups. "If each professional society were to develop their own materials, there would be so much material that chaos would result. Efforts should be part of a school system, curriculum-development group, or state group program. Thought should be given to how the materials will be utilized or combined with other societies. Coordination will build support and institutionalization of the product and some confidence that the work is going to pay off."

Sustaining Community Support for Science Education

The *National Science Education Standards* recognizes that a broad base of community support will bring about long-term improvement in science education. Professional societies should communicate with other community organizations that promote K-12 science education. Such organizations include parents, K-12 school districts, civic organizations, community colleges and universities, science-rich institutions, and industrial partners. Scientists, as individuals and representatives of agricultural professional societies, can play a role in science education reform.

This section is divided into three parts. First, strategies to link professional societies with communities are identified. Second, an excellent example of community involvement in implementation of the science standards is described. Finally, steps to sustain a science education program are summarized.

LINKING PROFESSIONAL SOCIETIES WITH COMMUNITIES

Societies can increase their impact on reform by exploring opportunities to link with organizations beyond the scientific community. Forum participants identified some ideas:

- form focus groups to help societies determine community needs. For example, one society expressed frustration over their lack of influence on proper nutrition information in schools;
- gain support for inquiry-based education from other agricultural organi-

BOX 3

The Impact of Scientists

Paul Williams, University of Wisconsin

"Every elementary teacher can amplify what you find exciting about science or what you do for 20 to 30 kids per year; over the [teacher's] lifetime, that translates into hundreds. If you are a high-school teacher, that increases to as many as 100 to 125 per year and thousands in a teacher's life. That's just part of the amplification system; imagine if you can get your messages and the excitement you feel for science to the teachers of the teachers. These are the professors of science education in schools and colleges of education. Each of them may teach from 30 to 130 teachers of science per year. Now we're talking about compound interest!"

Relationship of Scientist in Science Experience	Number of Children Influenced	
	1 year	30 years
• Scientist mentors student research	1	30
• Scientist visits school class	25	750
• Elementary teacher has science experience	25	750
• Secondary teacher has science experience	100	3,000
• Biology education professor who teaches		
— 20 elementary teachers per year for 30 years	15,000	450,000
— 10 secondary teachers per year for 30 years	30,000	900,000
• Coalition of educators and scientists influences 20 biology educators who each teach for 30 years		
— 20 elementary teachers per year	300,000	9,000,000
— 10 secondary teachers per year	600,000	18,000,000

SOURCE: Paul Williams, University of Wisconsin.

zations, such as the Cooperative Extension Service, 4-H Clubs, and Future Farmers of America;

- partner with local industry representatives to increase the support for science and technology education of students who could be employed by these industries in the future;
- involve local civic organizations, such as the Rotary Club or the Kiwanis Club, in activities such as math and science nights presented with the Parent-Teacher Associations (PTAs).

THE IMPACT OF COMMUNITY OUTREACH ON EDUCATION

Patricia Hoben
Minnesota Public Utilities Commission, Minneapolis

Communities that develop networks can disseminate resources to educators more broadly and efficiently. Patricia Hoben, research director at Minnesota Public Utilities Commission, described the dramatic results of having a communitywide outreach program in Minneapolis. The program maximized its impact on education by involving community institutions early in the process. The first step was to identify representatives of corporations, colleges, and universities who were involved in outreach. "These people were brought to the table to help develop local standards and plan for a revitalized science program in the Minneapolis public schools. At the same time, they became knowledgeable of activities that are responsive to teachers' needs. The group was awestruck at their potential power.

"Another aspect of this outreach is figuring out how to maximize the impact of a community institution. Each organization has invested their time and energy in developing a product, such as an in-service activity, curriculum supplement, field trip, or museum exhibit. The developers want as many people as possible to see and use it.

"We have gone to corporations and discussed how they could fund programs which are consistent with teachers' needs. What we are trying to do is find ways to alleviate the burden of individual teachers in finding resources by making these kinds of materials and activities known to the whole population. In this case, Minneapolis has 90 schools, 60 of which are elementary schools; we have 3,000 teachers and 44,000 students. Our goal is to give all of those individuals and institutions access to what's available through those institutions. Our community institutions are really receptive to working together as long as they can retain their autonomy and meet organizational objectives.

"Many other activities exist in the Twin Cities area outside of what has been discussed, such as school lectures, class visits, and adopt-a-school programs. There are summer institute offerings and individual resource centers outside the schools. It may be beneficial to coordinate all these activities. Our goal is to avoid creating an institution which would require funding from local supporters. This would create conflicts, and we would have to compete for resources. We see our role as more like a catalyst that does not get used up in the reaction. We're really at a very early stage of this process."

STEPS TO SUSTAIN A SCIENCE EDUCATION PROGRAM

Michael Klentschy
Superintendent, El Centro School District, El Centro, California

Michael Klentschy defined six points to sustain a science education program:

"1. The school board has to believe that science education reform is their highest priority. It is worth noting that these school-board members are usually elected and they have two goals: (1) to serve the community and (2) to get re-elected.
2. Develop a broad base of support in the community. As more people become involved in developing and implementing a program, community support will become stronger and the more likely the program will be sustained.
3. Get the buy-in from the teachers. Make sure they have a sense of ownership of the program, otherwise it's not going to be sustained.
4. Educate parents. They have different memories of school science. No longer are we giving out grades on lab notebooks or handing out textbooks. They need to understand these changes. You can do that by involving them in science nights. That builds tremendous support with your advisory councils, your PTAs, PTSAs, and all of the other groups that are involved.
5. Involve the media and promote the positive things going on in our schools.
6. Don't let your program get static; let the materials and curriculum evolve. Do that by having the science community and the education community work together in science education reform."

Future Roles for Professional Societies and Scientists

Many opportunities exist for professional societies to contribute individually or collectively to K-12 education. Each society should be committed to science education and work to implement the goals of the *National Science Education Standards.* In addition, teachers and schools need help to improve the teaching and learning process, including the tools and techniques that they use. Professional societies can use their creativity to enhance the teaching and learning of children.

UMBRELLA ORGANIZATION

The discussions in the forum clearly showed that every society was devoting more and more energy toward promoting educational materials and addressing K-12 issues. An explosion of programs is creating an information overload for K-12 classroom teachers and making distribution and dissemination difficult. Although individual members of societies have made individual contributions to science education reform, linking their activities would be more effective.

Participants suggested that information technologies will be important to effective communication. The World Wide Web might be a good way for societies to communicate with educators and other societies; more importantly, electronic connectivity will enhance access to information and new collaborations among teachers, scientists, and students.

The participants agreed that an umbrella organization was needed to provide leadership for moving the new science education standards forward. Various groups in the technical community are convening conferences to discuss education reform. For example, Sigma Xi had a conference in 1994 and published the proceedings: *Science, Educators, and the National Standards: Action at the Local Level.* A similar topic was the subject of a recent National Research Council report: *The Role of Scientists in the Professional Development of Science Teachers in 1996.* Scientists are interested in finding a logical group to coordinate the professional societies. Just who could commit to that responsibility is unclear. Representatives suggested that a strategic plan be developed to identify roles individual professional societies could play in an umbrella organization. Although some participants suggested that the National Research Council be a convener, Carla Carlson, former staff member of the Board on Agriculture, indicated that some opportunities might exist already to integrate various society activities. Participants pointed out that duplication of organizational efforts could be avoided by working with an established national group, such as the Coalition for Education about Environment, Food, Agriculture, and Renewable Resources (CEEFAR). Certainly, the umbrella organizational concept is a topic that deserves more discussion within each professional society.

Participants also discussed having regional or local gatherings. One of the benefits of having local meetings would be that scientists and educators could meet more easily and create connections that could be developed later.

RESOCIALIZATION OF ACADEME

In many cases, university scientists are interested in participating in K-12 education, but obstacles to gaining their widespread support remain. College and university faculty and administrators will need to coordinate their efforts in planning activities and course work for practicing and prospective teachers.

Incentives to involve scientists in K-12 educational reform should be considered. Some scientists might become involved, but in general, such activities are not valued at the university level; the requirements for promotion and tenure make it more difficult to get involved in reform activities. Educators point out that scientists are often reluctant to volunteer for K-12 classroom activities (see Box 4). Thus, incentives to encourage scientists to participate in K-12 science education should be discussed.

Scientists are concerned that university administrators do not encourage them to participate in K-12 education. Universities will need to address faculty rewards, particularly scholarships, awards, promotions, tenure, and publication opportunities. Based on the discussion, a suggestion was made that the professional societies could help to solve the problem of rewards by returning to their local universities to work with faculty administrators to develop rewards for faculty involvement in K-12 education. Paul Williams noted "that part of the reason that we're here today is to begin to wrestle with the resocialization of academe within the hierarchy of the tenure system, so that it is acceptable, in fact desirable, to encourage transitions." Many participants felt that the professional societies

BOX 4
Involving Scientists in K-12 Education

Kathy Scoggin, Minneapolis
Marcy Open School

"One of the best examples occured this fall. I was doing a river watch, and I was planning to take all the kids down the river and collect little critters and do a count of water quality. I decided that I would really like an aquatic entomologist to come with me because I didn't know enough. So, in my naiveté, I called the university, and said, `Would you like to come?' They said, `No, try this number.' So about 5 numbers later, I finally got a student who had just finished her graduate studies. She'd had a career as a physical therapist, her children had grown up, she went back to college, and she decided, "I really did always like insects; that's what I'm going to learn about this time." So instead of being a physical therapist anymore she decided to become an entomologist. She didn't have a job yet, so the professor suggested that I try her. This is after I'd been turned down a number of times. I called her, she said, `Sure, I'll come in.' It was the best thing I've done all year.

She brought more nets with her than I had, and she brought collecting cans; I learned things from her about the logistics of how I was going to do this with all of those kids and make it safe and meaningful. And she learned all about those kids because she couldn't believe that they'd be so excited and want to know about these critters—'What is that?' and 'Do you see this one doing that?' and 'Look, there's one on top of another one!' when they'd find little parasites. She was so excited she said she would come back and began suggesting other projects.

She learned a whole different thing about schools because she was there with kids, and I learned from her because I needed to know more about entomology to do a decent job. It wouldn't have happened if it hadn't been a one-to-one kind of thing; if we had just met and talked about it, it wouldn't have been the same. If you are interested in going a little bit beyond sharing your content knowledge, I think you need to try being there, seeing what it's like, volunteering for half a day."

should have a role in that process, but that the changes must come from within the university.

"I guess what I'm thinking is that our professional societies need to work out, within the context of their own cultures, some sort of operative structures that work for our particular cultures, that enable younger people with the desire and the zeal to do something for teachers in this context," Paul Williams continued. "As far as I'm concerned, this means encouraging them, letting them go, and changing the way we evaluate tenure on our university campuses so that they are recognized for the kind of contributions that they are making. Those are huge changes for academe to take; they're larger at some institutions than others; but they're all integral to this reform movement that we're talking about. I don't have the solution, but when I go back I know what I can do on my own turf. I think that your challenge, as representatives of your societies, is to ask how can you mobilize that notion in the individuals back in your own institutions. I think that the professional societies have an important role in validating the transitioning of faculty or or staff members, and in changing the structure of the universities and colleges so that they recognize this kind of interface as important to the whole system's future."

THE VALUE OF THE FORUM

Judith Williams summed up the value of the forum for participating teachers. "When I started teaching, it was my own little classroom . . . if we truly believe in our students and their outcomes and science education, it has got to become a partnership, and we've got to recognize each other's needs and what we can do together, or it's our students, our children, our grandchildren, our neighbors' children who will pay the price."

In speaking for participants, William DeLauder indicated that, "Several challenging and exciting things have been presented at this workshop, but they do not necessarily lead societies down the same path. It is obvious from the comments of the different groups that goals of cooperation, coordination, and communication are shared among participants. How best to achieve these is obviously the dilemma. It also is very important to recognize the tremendous implications of K-12 reform for higher education and all formal and informal educational endeavors. Scientific societies have been primarily research based. If societies are to have impact, then it must be recognized that the *National Science Education Standards* don't say anything about agriculture. You find "food" and "renewable resources" twice. But their absence provides societies with the opportunity to address how agricultural sciences affect the application when we get down into the frameworks and the curricula. That is the challenge." Agricultural professional societies have performed a broad range of services in higher education, adult learning, and provision of technologies. Societies need to consider how they can enhance early student learning in science and emphasize K-12 education.

APPENDIXES

APPENDIX A
Speaker Biographies

William B. DeLauder, a scientist, educator and administrator, became the eighth president of Delaware State University on July 1, 1987. He manages a campus of more than 3,300 students and 175 faculty. Prior to assuming this post, he served with distinction as dean of the College of Arts and Sciences at North Carolina Agricultural and Technical State University for 6 years. During 16 years at the university, DeLauder also served as associate professor of chemistry, acting chair of chemistry, and professor and chair of chemistry. He received a B.S. degree from Morgan State College and a Ph.D. degree in physical chemistry from Wayne State University. He subsequently conducted research in physical biochemistry at the Centre de Biophysique Moleculaire du C.N.R.S. in France as a postdoctoral fellow from 1969 to 1971. His research on the physical properties of macromolecular systems and on the florescence properties of proteins has been published in leading scientific journals. In 1990 he was appointed to the National Advisory Council of the National Institute of General Medical Sciences of the National Institutes of Health.

Patricia Hoben is Assistant Director of The Bakken, a museum and library in Minneapolis, Minnesota that focuses on applications of electricity and magnetism in life. Dr. Hoben is also Principal Investigator and Co-Director of a National Science Foundation grant to stimulate reforms in the Minneapolis public-school science program and promote local partnerships among K-12 teachers and scientists in higher education, industry, and science museums. Dr. Hoben received her Ph.D. degree in molecular physics and biochemistry from Yale University. She was a science policy advisor to the Assistant Secretary of Health and at the U.S. Congress Office of Technology Assessment, and directed the

Howard Hughes Medical Institute's precollege and public science education grants program.

Michael P. Klentschy, currently the superintendent of schools of the El Centro School District in El Centro, California, has served as the associate superintendent of instruction for the Pasadena Unified School District, in addition to serving as teacher, principal, federal programs administrator, and administrator for elementary instruction for the Los Angeles School District. Dr. Klentschy received his Ph.D. degree from the University of California, Los Angeles, in educational policy studies and served as an instructor in the Graduate School of Education there for 13 years. Working with staff from the California Institute of Technology, he also served as co-principal investigator on several elementary science education projects for the National Science Foundation Initiatives program.

Harold Pratt has had extensive curriculum development experience at the local and national levels. He served for 23 years as the science coordinator for the largest school district in Colorado and continued to oversee the operation of the Science Department and serve as executive director of science and technology for the next 5 years and later as the executive director of curriculum. In 1986, Pratt received a grant from the National Science Foundation to develop the Middle School Life Science Project. Working with a co-director and a team of teachers, Mr. Pratt developed a year-long course containing laboratory activities integrated into the text. The program, which incorporates cooperative learning, was piloted, field tested, and later published. Mr. Pratt served on the National Research Council National Science Education Standards Project and in January 1995 became the project director for the revision of Science for Life and Living for Biological Sciences Curriculum Study. Since May 1996 he has served as Director on K-12 Policy and Practice in the Center for Science, Mathematics, and Engineering Education at the National Research Council.

Kathy D. Scoggin has been a teacher in Minneapolis, Minnesota since 1975 and has been actively involved in the Minneapolis early-education science and environmental and conservation programs. She served as a facilitator/leader for Project WILD, an international, interdisciplinary conservation and environmental education program. For Project WILD, Ms. Scoggin managed workshops for teachers, instructing the teachers in innovative techniques for teaching basic skills in environmental and conservation sciences and in conducting hands-on activities that enhance student learning in all subject and skill areas. Ms. Scoggin is also a member of advisory committees for Science Centrum, University of Minnesota; the Minnesota Department of Natural Resources; Minnesota Landscape Arboretum; the Bell Museum of Natural History; and the Science Museum of Minnesota.

Jan Tuomi is currently the director of the National Research Council Regional Initiatives in Science Education (RISE) project. Previously, she was the director of Outreach for the National Science Resources Center, a K-12 science education improvement program founded by the National Academy of Sciences and the Smithsonian Institution. Ms. Tuomi's involvement in education includes 24 years as an elementary teacher and leader of professional-development activities for teachers. She received a B.S. degree in education from the University of the Pacific, taught in California, Colorado, New Mexico, and Mexico. Ms. Tuomi views optimizing the involvement of scientists in improving science education as her focus of interest.

Judith Williams, a biological sciences teacher, teaches high-school biology, applied biology, and advanced biology in Central City, Nebraska. Ms. Williams received her B.S. and M.S. degrees from the University of Nebraska at Kearney. She has been actively involved in the development of the Mathematics/Science Frameworks for Nebraska schools and was selected by the Nebraska Department of Education to be part of a writing team to develop standards for secondary science teachers and to be a professional-development team leader. She has also conducted the Nebraska Mathematics/Science Frameworks Workshops for K-12 teachers. Earlier this year, Ms. Williams presented a paper at the 1995 Invitational Conference on Systematic Reform in Science Education sponsored jointly by the National Science Foundation and the U.S. Department of Education.

Paul H. Williams, a professor in the Department of Plant Pathology at the University of Wisconsin-Madison, is the developer of the rapid-cycling "fast plants," used extensively as models for research in a wide range of biological applications and as organisms for hands-on exploratory learning. The Wisconsin Fast Plants™ and Bottle Biology™ programs are derivatives of his research and have been used at all levels of education—from preschool through college. Supported by the National Science Foundation and the W. K. Kellogg Foundation AgriScience Institute, these programs have been demonstrated to more than 40,000 teachers in training workshops throughout the United States, Canada, Australia, the United Kingdom, and other countries. More than 2 million students use Wisconsin Fast Plants annually. Dr. Williams received his Ph.D. degree in plant pathology from the University of Wisconsin-Madison.

APPENDIX B

Meeting Agenda

Agriculture's Role in K-12 Education
November 17 and 18, 1995

National Academies of Sciences and Engineering Beckman Center, Irvine, California

FRIDAY, NOVEMBER 17

Introduction
William DeLauder, Board on Agriculture member and President, Delaware State University

History and Context of the National Science Education Standards *Report*
Harold Pratt, Visiting Scholar, Biological Sciences Curriculum Study, and Senior Program Officer, National Science Education Standards Project

Q&A Session

Science as Inquiry
Paul Williams, Professor of Plant Pathology, University of Wisconsin

Workshop: Experiencing Fast Plants™
Paul Williams

Lunch break

Agenda and Format for the Afternoon Panel and Discussion Groups
William DeLauder

A Scientist's Role in Education: A Teacher and Educator Panel
Patricia Hoben, Science Policy and Education Consultant, Minneapolis, MN
Michael Klentschy, Superintendent, El Centro School District, El Centro, CA

Kathy Scoggin, Marcy Open School, Minneapolis, MN
Judy Williams, Central City Public Schools, Central City, NE

Moderator: Jan Tuomi, Center for Science, Mathematics, and Engineering Education, National Research Council

- Curriculum decisions and design
- Teacher education
- Coordination of resources
- Developing and sustaining community support

Q&A Session with Teacher and Educator Panel

Breakout Discussion Sessions (discussion questions will be distributed in the registration packet)

- *Facilitators:* Norman R. Scott, Cornell University, and Judy Williams; *Recorder:* Roland Otto, Society for Industrial Microbiology
- *Facilitators:* Susan K. Harlander, The Pillsbury Company, and Patricia Hoben; *Recorder:* Robert Zimbelman, American Society of Animal Science
- *Facilitators:* Richard R. Harwood, Michigan State University, and Paul Williams; *Recorder:* Janet L. Bokemeyer, Rural Sociology Society
- *Facilitators:* George E. Seidel, Jr., Colorado State University and Harold Pratt; *Recorder:* Deborah Neher, Ecological Society of America

Agenda for Saturday Plenary Session Summaries

SATURDAY, NOVEMBER 18

Plenary Session: Summaries of the Four Discussion Groups by Session Recorders

Synthesis: Conclusions and priorities
William DeLauder

Q&AClosing Comments
William DeLauder

Professional scientific society representatives conduct their agenda on science education issues, common concerns, other issues

APPENDIX

C

Forum Participants

Ken Anderson
American Society for Microbiology
California State University
Department of Biology and Microbiology
Los Angeles, CA

Walter J. Armbruster
Farm Foundation
Oak Brook, IL

Robert F. Barnes
American Society of Agronomy
Crop Science Society of America
Soil Science Society of America
Madison, WI

James L. Barrentine
Weed Science Society of America
Dow Elanco
Cordova, TN

Daniel P. Bartell
American Association of State Colleges of Agriculture & Renewable Resources
California State University, Fresno
School of Agricultural Sciences and Technology
Fresno, CA

Janet L. Bokemeier
Rural Sociology Society
Michigan State University
Department of Sociology
East Lansing, MI

Cathy Bridwell
Agricultural Communicators in Education
U.S. Department of Agriculture
Extension Service
Washington, DC

Leonard Bull
American Dairy Science Association
North Carolina State University
Raleigh, NC

Rueben Buse
American Agricultural Economics Association
University of Wisconsin
Department of Agriculture Economics
Madison, WI

Frank Busta
Institute of Food Technologists
University of Minnesota
Food Science and Nutrition
St. Paul, MN

Lucas Calpouzos
Project Food, Land & People
Dean Emeritus of Agriculture
California State University
Chico, CA

C. Lee Campbell
American Phytopathological Society
North Carolina State University
Department of Plant Pathology
Raleigh, NC

Vernon Cardwell
Coalition for Education about Environment, Food, Agriculture, and Renewable Resources (CEEFAR)
American Association for the Advancement of Science
American Society of Agronomy
Crop Science Society of America
Soil Science Society of America
University of Minnesota
Agronomy Department
St. Paul, MN

Neville P. Clarke
National Association of State Universities & Land-Grant Colleges (NASULGC)
Texas A&M University System
Southern Agricultural Experiment Station Directors
College Station, TX

Darleen A. DeMason
Plant Sciences Bulletin
Botanical Society of America
University of California, Riverside
Department of Botany & Plant Sciences
Riverside, CA

Dean D. Duxbury
Institute of Food Technologists
The Society for Food Science and Technology
Chicago, IL

Vernon R. Eidman
American Agricultural Economics Association
University of Minnesota
Department of Applied Economics
St. Paul, MN

Ralph A. Ernst
Poultry Science Association
World's Poultry Science Association
University of California, Davis
Department of Avian Sciences
Davis, CA

Robert Fridley
American Society of Agricultural Engineers
University of California, Davis
Agriculture and Environmental Science
Dean's Office
Davis, CA

Bill Frost
Society for Range Management
UC Cooperative Extension
Placerville, CA

Sharon Curtis Granskog
American Veterinary Medical Association
Schaumburg, IL

Tom Helm
Casey Family Foundation
Headquarters
Seattle, WA

Walter A. Hill
Association of Research Directors
Tuskegee University
Tuskegee, AL

Geza Ifju
Society of Wood Science and Technology
Forest Products Research Society
Virginia Polytechnic Institute and State University
Department of Forest Products
Blacksburg, VA

Joseph Kunsman, Jr.
National Association of State Universities & Land-Grant Colleges (NASULGC)
Washington, DC

Shelli Lamb
Soil and Water Conservation Society
Riverside Corona Resource Conservation District
Riverside, CA

Peggy G. Lemaux
American Society of Plant Physiologists
University of California at Berkeley
Berkeley, CA

Deborah Neher
Ecological Society of America
University of Toledo
Department of Biology
Toledo, OH

Mortimer H. Neufville
Association of Research Directors
University of Maryland, Eastern Shore
Early Childhood Research Center
Princess Anne, MD

Christian Oseto
Entomological Society of America
Purdue University
Department of Entomology
West Lafayette, IN

Roland Otto
Society for Industrial Microbiology
Executive Director, California Science Project (Interim) and Center for Science and Engineering Education
Lawrence Berkeley National Laboratory
Berkeley, CA

Warren Schwecke
Council for Agricultural Science and Technology
General Mills, Inc.
Minneapolis, MN

Ham Shirvani
Graduate Education and Research
Queens College, CUNY
Flushing, NY

Charles F. Shoemaker
American Chemical Society
University of California, Davis
Department of Food Science & Technology
Davis, CA

James L. Starr
Society of Nematologists
Texas A&M University
Department of Plant Pathology & Microbiology
College Station, TX

Richard Stuckey
Council for Agricultural Science and Technology
Ames, IA

Bert Swanson
American Society for Horticultural Science
University of Minnesota
Department of Horticultural Science
St. Paul, MN

Teresa Thomas
American Society for Microbiology
Southwestern Community College
California Biotechnology Education Consortium
Chula Vista, CA

Robert Vance
American Meat Science Association
California Polytechnic State University
Food Science and Nutrition Department
San Luis Obispo, CA

Conrad J. Weiser
American Society for Horticultural Science
Oregon State University
College of Agricultural Sciences
Corvallis, OR

Coe Williams
Wisconsin Fast Plants
Madison, WI

Elizabeth A. Wolanyk
Agriculture in the Classroom
U.S. Department of Agriculture
Office of the Secretary
Washington, DC

Sheri Zidenberg-Cherr
American Institute of Nutrition
University of California at Davis
Department of Nutrition
Davis, CA

Robert Zimbelman
American Society of Animal Science
Federation of American Societies of Food Animal Sciences
Bethesda, MD 20814